The One Hour Ecologist

S J Loran

Grosvenor House
Publishing Limited

All rights reserved
Copyright © S J Loran, 2025

The right of S J Loran to be identified as the author of this
work has been asserted in accordance with Section 78
of the Copyright, Designs and Patents Act 1988

The book cover is copyright to S J Loran
Cover design by Lulu Da Costa

This book is published by
Grosvenor House Publishing Ltd
Link House
140 The Broadway, Tolworth, Surrey, KT6 7HT.
www.grosvenorhousepublishing.co.uk

This book is sold subject to the conditions that it shall not, by way of
trade or otherwise, be lent, resold, hired out or otherwise circulated
without the author's or publisher's prior consent in any form of
binding or cover other than that in which it is published and
without a similar condition including this condition being
imposed on the subsequent purchaser.

A CIP record for this book
is available from the British Library

ISBN 978-1-83615-069-5

For Ozzy and Frank

Introduction

Ecology in its simplest definition is the science of the relationships among and between living organisms – and that includes humans and how we interact with the world around us. We have an hour, which is not long to skim the surface of an enormous subject. We will look at what ecologists do, and how we can all play our part, because we are all absolutely involved in the ecology of the Earth every day of our lives.

Most of us have more than a vague idea that the stuff we use everyday comes from somewhere else. We know it was created in a factory, or many factories, using different components that were sourced from all over the world. I'm writing this in the UK. I sit at my laptop wearing a t-shirt and jeans in full knowledge that every component of my clothing is from another country; cotton doesn't grow in the UK, it likes a hotter drier climate, India say, or the Southern United States. I know that the zip and the buttons are made from Nickel. Nickel is mined in Indonesia, the Philippines, Russia, Australia, and Brazil. So, what is very ordinary stuff for us has a long chain of components that are mined or grown all over the world. When we buy that product, we rarely give the provenance of its constituent parts much thought.

As an ecologist we would be thinking about where the mining or the growing was happening? And what was there before this process began? And what has been destroyed, displaced, or disrupted by this process coming into being? So you're an ecologist who is trying to save a particular species of animal from extinction and underneath your animals last remaining habitat are minerals and ores that will be valuable to industry, how would you go about trying to stop that mining from happening?

Mining is big business, ecology isn't, mining makes billions of dollars, ecology doesn't. The industry that will benefit from the sale of those resources rarely has to answer for the land and the habitats that it destroys along the way.

We are as a global society so profoundly signed up to our right to acquire anything that we can legally obtain, and we are very rarely informed about the products consequences. And so, the question then falls to us, the consumers of those products, are we prepared to lose that animal from existence to create a huge mine so that those minerals can be extracted in order that we can continue to sustain our current lifestyle, computers, phones, cars, clothes, airlines? Because these are the ecological choices that we as consumers have the power to make, it's our hard-earned cash that is paying for those mines to exist.

Some people already make huge efforts for the planet, I have a friend Susie who is in the group Extinction Rebellion. She would do everything in her power as an individual to save that animal and its habitat, so whatever was underneath that forest that threatened the animal with extinction she would go out of her way to avoid buying in the future. But it isn't easy to do this. What if the product is oil? We all need to get to work. What if that product is Coltan? We all need to keep in touch on our phones. It's incredibly complex and most of us are guilty to some degree in being participant in the destruction of a previously beautiful habitat. We just are. There is little we can do to avoid this truth. Most of us just aren't as far down the road as Susie, if you are, then most likely you're reading the wrong book!

Here is Susie's life in a nutshell. She doesn't fly, she's given up her car, she eats only vegan mostly organic food, she banks ethically and has ethical investments from companies that have divested from fossil fuels (meaning they won't put their money into any company that is related to the oil and gas industry – who often extract oil from under forests and wild places), she only buys second hand books, clothes and furniture, she only uses ecological cleaning products and those bars of shampoo and conditioner, she's pretty much plastic free, she even tries to keep the heating off for most of winter, this is what good for the planet really looks like. Susie is seriously impressive and by comparison I don't stack up nearly so well. Susie is single and has no pets or children. She lives in a shared flat. She does have a mobile phone that's her one bad thing that she just cannot account for, so even Susie, you'll be glad to know, might have unknowingly participated in the destruction of a pristine wilderness.

I am your writer, I'm married (to a meat-eating fellow who has a bad habit of not buying ecological cleaning materials when I send him to a shop). I have put two (I have to say rather nice) children on the planet and I have two carnivorous pets, so from the get-go I would find it impossible to achieve Susie's standard of lifestyle impact reduction. But I've given up feeling like a hopeless environmental criminal around her. I aspire to be a little bit more like her and a little bit less like me, but it's impossible for me to achieve the Susie gold standard of being a vibrant western human who is doing least damage to the planet during her time here.

I did the WWF ecological footprint calculator (https://footprint.wwf.org.uk) – and I come in at a very average 7 tons of carbon a year – and I also don't fly, I use trains and buses where possible, I buy all my gear in charity shops, and I don't eat meat (but I do occasionally eat fish), I recycle like a mad woman and I get a veg box and grow some of my own. So how do I score so averagely? Well, I live in a house, have kids, have pets, have car, have a meat eater husband. The WWF footprint calculator looks at your household, not just you. I can't see what I can do about some of those things that drag my score, I can't not have the kids, or get rid of the pets (both are rescues). I'm not sure I can force my husband to give up meat. I have elderly parents and work nights, so the car is important. I found it rather dispiriting though that my constant effort to tread more lightly on the earth brings me such little recognition at WWF. Here is a recent example of how good I can be from early June 2024 – my daughter lives in Edinburgh to fly was £23.50 each way, to take the train was £206 return.

People like me would happily endure the long train journey because its more environmentally sound to do so, but I can't afford to pay £206 for a three-day break.

What did I do, well, I just didn't go to Edinburgh. If I had been going for a week, I'd probably have got the bus, but I only had a few days free, it just didn't work out because I wasn't prepared to fly. So, I am good. But my ecological footprint is still disappointing.

I wonder how many of you when you read about Susie would respond like another friend of mine Maggie. Her response to Susie's lifestyle is along the lines of, 'what's the point of being alive?'.

Susie lives in Lewes in Sussex, it's one of those liberal towns with a very activist community and quite literally Susie is having a pretty good time… 'hot yoga Mondays', 'drumming workshop Tuesdays' 'book group Wednesdays' she likes hiking and takes the train to hiking destinations, she likes film (they have a very cool cinema in Lewes), she enjoys meeting all of her XR buddies and planning their next move (I don't belong to XR in case you're wondering, it's just not for me, thanks) she's in a theatre group, she's in a choir, she volunteers in a nature garden, she's a wild swimmer, she goes to a flamenco class. I could see Maggie's brain whirring as I delivered Susie's hectic schedule, and when I list her weekly activities, I'm thinking she has a much more exciting existence than I ever have! Who knew, you don't need to destroy the planet to gather in your community and have a good time.

But Maggie is having none of it. She shakes her head and takes a new tack, that all of Susie's efforts to not drive or fly are 'a drop in the pissing ocean'.

And it really can feel like that can't it? I do find myself thinking sometimes, but if I don't fly the plane will take off anyway. If I don't buy that item, it will be sold to someone else. That fish is dead now anyway, so I might as well buy it and eat it. It sometime feels difficult to make the right choices and to feel that these small gestures have any real meaning to them. But when your small gesture meets with hundreds of other small gestures it will make a difference. And if you live somewhere like Lewes you can see it happening in real time, a community of like minded folk gather and share and enjoy each other.

I always recall that quote from the Dalai Llama when I'm feeling powerless, 'if you think you're too small to make a difference try spending the night with a mosquito' and when I think of that I get back on track to what is important to me, it's my performance as a concerned human being living on an ailing planet.

Susie knows that giving up her car isn't going to make a huge difference to anything, initially it was a symbolic gesture, but it has led me (and maybe others) to fantasize about a time in the future when I too might be able to live car free, it is an aspiration now, I can see that for her it works and I prefer taking a train or bus anyway, so I'm there in spirit, we just need intelligence at the top that will make sure the train is a meaningfully cost effective alternative to driving, or flying.

Our efforts will only be met by big business or come into government policy when their bottom line begins to show that we are making different more conscious choices, all the time we go along with the business as usual that's what we'll see. What I hope as you read this book is that you will see how implicated we consumers are, it really is us, the poorer, the middling and the richer that are making all of these choices that will either feed into the planets resources being endlessly overused, or not, it's up to us, the wage slaves and workers to decide the direction of travel, so shall we begin.

CHAPTER ONE

But what is ecology and who are ecologists?

In its most simplified definition, we live on Earth. Earth is a diverse lump of rock that contains many different climate zones and each of these climate zones (ecologists call them Biomes) has different plants, animals and insects (ecologists call this biota) and water needs (ecologists call this hydrology).

What must underpin any discussion about ecology is the simple fact that the earth evolved over billions of years. And during its many periods of being literally on fire for a million years and then literally freezing for a million years it broke down layers of its rocky surface enough to create something like soil. And eventually with enough water and sunlight, plants came. They grew and lived and died and that process of them rotting down back into the rocky soil helped to create richer soil and more plants grew and because plants photosynthesise which means they absorb carbon dioxide from the atmosphere they use this with sunlight and water to create glucose energy for growth and because the waste product of this process is oxygen this created an atmosphere and with enough oxygen eventually animal life happened.

And we currently live on a planet that has the right atmosphere and temperature to support an extraordinary variety of life. But we know it hasn't always been this way. And we know that its continued existence as a habitable lump of rock depends on these plants and animals continuing to be able do what they do because they regulate our water, our atmosphere and provide our food, so we need all this ecology to be working on our side.

Here are some examples of Biomes; tropical rainforest (Brazil), arctic tundra (Greenland), maritime temperate (UK), Mediterranean scrub (Spain), equatorial savannah (Tanzania), Ocean (globally).

Each of these biomes might contain many separate ecosystems and is composed of an in inestimable number of living organisms. These might be plant or animal and range from the simplest single cell organisms like amoeba to the most complex life forms like rhinos or birds.

Ecology is also about ecosystems, the environments in which many different types of organism exist and how they all contribute to the overall health of the earth system.

The types of environments that ecologists might call an ecosystem are a forest, a woodland, a river, an ocean – each of these areas has a diverse number of organisms or species that are active although not always limited to that place. For instance, a fox might be part of a woodland ecosystem, but foxes don't just remain in the woodland, they will hunt far more widely. The same is true of birds and many flying insects, so ecosystems can be quite fluid with their inhabitants, but, if they are providing the right plants, soils, water and nutrients then the creatures who thrive in those conditions will always be present.

The reason that we are interested in ecosystems is that when they function as they should they have a good range of organisms; enough plants, animals, insects, microorganisms, soils, atmospheric gases, and water to continue to be healthy and survive. And healthy ecosystems give out oxygen and make a healthy planet.

When an ecosystem is unhealthy it generally means that some component has been removed and there is a knock-on effect. This could be a drought, which means that without enough water plants will die, and animals might succumb to dehydration. Or something has come into that ecosystem that is creating a problem, such as a factory discharging its waste containing toxins into a river and killing off the fish. In situations like this the ecosystem can actually begin to emit carbon dioxide instead of oxygen and then it becomes an emitter of greenhouse gases, which are warming the planet.

Sometimes an ecosystem is destroyed in its entirety. Imagine how many forests were felled to create a London or a New York. And you might say yeah but that was a very long time ago, and it was, however towns are still being created today, or added to. We all know the term urban sprawl, and if that sprawl happens to be where nature used to live then it will have impacted and probably destroyed that ecosystem.

Of course, it isn't just for the building of cities that ecosystems are destroyed. Beyond human infrastructure (homes, roads, factories, airports, railways – the built environment), we use vast swathes of the earths surface to feed ourselves. Our other means of destroying fragile ecosystems is for mining resources, be that for petroleum production, coal, metals, and increasingly mineral ores like coltan, tungsten, cobalt and copper. As it stands, we have as a species annexed 14% of the earth's surface for human purposes. 14% doesn't sound overly dramatic until you remember that 71% is water. So basically, we've had half the land, the only parts that humans haven't annexed are mountains, deserts and forests (which are under constant threat from humans).

Many ecologists work in industry where they are tasked with identifying species of ecological importance also called keystone species before any of these destructive human activities can take place, or identifying keystone or endangered species as a reason that the destructive human activity cannot go ahead.

The track record for success however isn't as impressive as we might like to hope, especially if the destructive human activity is in a country with poor environmental laws, likely this is also a poor country, and with that comes greater need to engage with any activity that might increase wealth, even if it means destroying precious habitat along the way.

An example of this is the case of lowland gorillas in the Virunga National Park in Democratic Republic of Congo, they are increasingly coming into conflict with mining companies who have found cobalt and coltan underneath the national park. To mine those minerals would destroy one of the last refuges of this beautiful intelligent and charismatic creature. For a poor country with a huge population, the prospect of exploiting that mineral wealth is tantalising. As you can imagine the choices are never clear cut, there are valid arguments on both sides, to deny a country prosperity for its people feels wrong, to deny the lowland gorilla a home and a continuation of existence as a species on this planet feels wrong. What can be done? Ecologists are the people often tasked with finding a solution to situations like this, recognizing human needs but trying to manage those without destroying the environment and its flora and fauna.

Keystone species are animals or plants that have a high impact within the ecosystem that they live, even though they might not be the dominant creature/plant in terms of numbers. An example is the Oak tree, it won't be the most populous tree in the wood, but it provides the most 'ecosystem services' (which means that it does stuff that benefits the planet, and animals and ultimately us). It does water cycling, drawing up huge amounts of water through its root systems which are then expired into the atmosphere after photosynthesis. It breathes out oxygen as a by product which we in turn breathe.

It provides the most shade from its canopy. It provides the most habitat and food for insects and birds. It is estimated that a mature Oak will support over 2,000 creatures or plants (according to The Woodland Trust) and therefore if that Oak was lost the impact on all the other creatures in that ecosystem would be dramatic even though many of those creatures do not directly draw on the Oak for survival but rather draw of the other creatures the Oak supports.

Chapter Two

A whizz through the ecology of a forest/woodland

Primary, virgin or old growth forest is forest that has existed before man has intervened in its planting, so every plant in that forest belongs to that area, no planting has been introduced by humans.

The Amazon is a good example of virgin rainforest. It is thousands of years old and has not been planted by humans. It is the most biodiverse ecosystem on earth containing hundreds of different species of trees, plants, fungi, animal, bird and insect. We frequently hear it referred to as the lungs of the earth, and most of us are aware that due to its sheer size it absorbs more carbon from the atmosphere than any other forested area on Earth. We refer to this as a carbon sink. The area of Brazilian rainforest lost to logging, mining, soya production and beef farming has reached over 17% with a further 17% degraded by change of land use. This is approaching the critical threshold that some scientists have set at 25%, at which point the forest ecosystem could collapse. If this happens the forest would become a dry scrubby savannah at which point instead of absorbing carbon from the atmosphere and releasing oxygen it would actually begin to release carbon dioxide and this would contribute to the heating of the planet.

Secondary, succession or new growth forest is forest that has been planted and is managed by humans.

Forests and woodland both describe areas covered by trees, forests tend to be larger and generally have more coverage by trees. Woodlands can be very small areas and tend to be less dense, allowing more light though the canopy.

Kielder Forest in Northumbria, England is an example of a secondary forest. It was planted in the1920's purely for the supply of timber in the UK for various uses such as paper, furniture, fuel etc. These types of forest contain much less biodiversity than the Amazon, generally a single tree species is planted, (which ecologists call monoculture), and often this tree is not native to the area in which it is planted, furthermore if it is from the pine family (fast growing, good for timber) which have very thin leaves, it is less useful at absorbing carbon dioxide than the broader leaved native English trees such as oaks beech and hazel. Because of this it will not support insects as well as an old growth forest that has evolved over hundreds of years with its own niche species of creatures. Even so, Kielder Forest has become an important ecosystem despite its monoculture tree planting. Other native plants have found a place to flourish amongst the trees and they have in turn attracted insects such as pollinators, and this will then have attracted insect eating birds and reptiles and bats and mammals and so on.

But trees aren't the only element in a forest or woodland ecosystem.

Soil

Without being melodramatic it could be said that soil is the stuff that underpins all of life on earth. No soil, no food, no life, not for us, or for anything else that lives on land, you, me the caribou and the buzzard are all dependent on soil because we eat either vegetables or meat from animals that ate grass and seeds.

Soil is the thin layer that sits above the earths rocky surface, it consists of broken-down bits of rock and sediments that have accumulated over millions of years, decayed bodies of insects and creatures, decayed leaf matter from fallen trees, other decayed plant matter and a whole lot of poo. Soil, as unglamorous as it might seem is a living history lesson, in the untouched soils of a woodland there is probably several hundred years of all of the above.

It also contains air and water and microorganisms and bacteria; it is estimated that soil contains over six billion microorganisms in a single teaspoon. Its hardly surprising therefore that it is considered the most biodiverse 'system' on earth.

Vast amounts of fungi or mycelium live in the soil, scientists have recently discovered that communication between trees happens along the networks that mycelium provide, they are like the phone line of the forest enabling trees to talk to each other. This was a stunning discovery that provides us with evidence of the intelligence that is found in nature, I'd love to know what they talk about!

Millions of insects live in soil, and each one will contribute to the soil's fertility.

The soil contains minerals like nitrogen and phosphorous and potassium and manganese. In this way we can see that some of the minerals in the soil are minerals that we also need to keep our bodies healthy, and thus a healthy soil produces a healthy plant, which we then consume and uptake the minerals from.

Soil science is extraordinary, and I recommend an evening lying on the sofa treats to hand and watching YouTube videos on this very subject (suggestions at end of book). I did this and discovered even Hollywood has made a film about how degraded soil emits carbon dioxide but healthy soil sucks CO_2 out of the atmosphere. Kiss the Ground (2020) is narrated and produced by Woody Harrelson. If Hollywood understands the absolute necessity of healthy soil, then we owe it to ourselves, don't we?

And once you appreciate the reality of what binds this earth, that it is an endless process of birth, death, eating and pooing you begin to appreciate how the current paradigm has definitely got overly complex.

Insects

Every insect within the forest is there performing its basic ecological function to survive and reproduce in order to perpetuate its species.

In order to survive most insects need to feed, whether they are worms that eat plant litter, bees, wasps and flies that feast on nectar, butterflies that feed on fruit, beetles that eat bark or beetles that eat poo or insects that eat other insects, they are all necessary for the function of the forest.

You can imagine that each forest will have many thousands of insects, and this is necessary, when you consider the tiny contribution that each tiny insect makes. It will need to be a species that reproduces in hundreds to have a positive impact on the ecosystem. Sadly, we are seeing decline in many of our woodland species, some birds and insects are reaching critically endangered status across the UK. And only time will tell how much these losses will contribute to the eventual failure of the ecosystem in which they are evolved to live. We do live in challenging times for nature, and this will have an impact on our own health and well being as time goes on.

Let's imagine losing the beetles that eat poo. Dung beetles are a good example because their numbers are in serious decline across the globe. The dung beetle as the name suggests is a poo eater, and they help to break down poo and this enables it to more rapidly nutrient the soil. Without them the process is disrupted; the soil composition might change and that in turn can impact its fertility or weaken the plants that grow in that soil making them more susceptible to disease. The other impact that results from loss of dung beetle numbers is an increase in flies. Flies are also attracted to poo and as they land they poo their own bacteria laden faeces, this then ends up in the grasses and soil and is consumed by other animals who will get infected as a result, this might be wild animals but it's often livestock. With livestock we can draw a direct threat to humans from the loss of a dung beetle. When farm animals are culled due to disease it pushes up the price of that meat. Whilst it might only be a little seemingly insignificant beetle, as we can see, it can have a very direct impact on our lives.

And this is what ecology is about, value judgments and having conversations about how you feel about creatures and their lives. If you don't feel especially compassionate about the loss of a beetle species, you wouldn't be alone. It has long been recognised in wildlife conservation that it's easier to raise money by selling the imperilled status of a larger more easily recognised species, biologists have even created the term 'charismatic megafauna' to describe those creatures that we all know; the gorilla, panda, tiger, elephant, lion, giraffe etc Its rather embarrassing to discover that allegedly we are more likely to dig deep into our pockets for a panda or gorilla because something about them reminds us of us.

Here is a basic food web in which you can see the interdependence of each organism in an ecosystem. Trees need insects for pollination = small mammals like deer and monkeys need trees for food and habitat = jaguars need deer and monkeys for food = beetles and worms need jaguar dung for food = soil needs beetles and worms for breaking down jaguar dung into easily accessible nutrition = birds need beetles and worms for food. All of these organisms need water, oxygen or carbon dioxide that they draw from the atmosphere.

Therefore, the ecologist can try and raise money for the charismatic megafauna when the beetle is at stake because through the food web that jaguar is as dependent on the beetle as every other part of the system, just at several removes. Ecological thinking therefore acknowledges that every small part of a system is necessary for the system as a whole to function.

Chapter Three

How we have an impact on a forest

We're all consumers of products that use wood, furniture being the most obvious, but how often do you pick up a free newspaper on the train, and discard it before you get off? I do it almost without thinking and then feel guilty about it when my conscience finally kicks in. At that moment I vow never to pick up another free newspaper from the rack, but only to read the second-hand ones already littering the train – there is a school of thought that suggests that the embedded energy (that is the energy used in the extraction, processing, manufacturing etc of an item) is taken on by the person who first purchases it. So, if I buy a new car the embedded energy of that vehicle will become part of my ecological footprint. If I sell that car on, the embedded energy has already been taken account of by me, so the second owner only has the daily inputs the oil and petrol as part of their ecological footprint. The good news here of course is that by buying second hand you are relieving yourself of the burden of an enormous ecological footprint (well, the WWF might not agree but burden of first purchase is a thing in some ecological footprint analyses).

Whilst it is true that most of us get our news from the internet these days, papers still exist and readership is still strong, the annual figures for the Metro taken each January confirm that almost a million papers are consumed per month, that is just one paper, in just one country, it is estimated that 4 million hectares of forest are lost annually just for paper. Fifty thousand trees are cut down in the UK each year just for Christmas wrapping paper (this is according to the UK government website in 2022. I'll leave it for you to decide therefore whether the statistic is robust enough that forever after your children shall receive their gifts wrapped in tea towels!). These figures are mind boggling, but just know, paper impacts the earths ecology, and most of the time the paper, which was once a tree, has a very finite level of usefulness for our lives, we read and discard, we rip open and discard, all those happy to exist trees die for such ephemeral human activities. And the deal is, do we really need to use so much paper, would Christmas really be cancelled if presents weren't lavishly gift wrapped, could we not find some other more sustainable way to disguise our gifts, this is a business opportunity for someone, come on, get your thinking caps on!

Let us fly through an environmental cradle to grave assessment of a newspaper.

1. Sapling trees are planted in a managed plantation, such as Kielder Forest in Northumbria.
2. Over the next ten to fifteen years the trees grow to a width that is suitable for use in paper industry.
3. Trees are chopped down using heavy machinery.

4. Trees are transported to a mill where the bark will be removed, and the finished product becomes timber.
5. The timber is cut into small pieces and boiled with chemicals until it forms a bulky pulp.
6. The pulp is fed into machinery that squeezes it into the flat shape of paper.
7. The paper is transported to a printers.
8. The paper is printed using petroleum or soy or water based inks.
9. Papers are distributed across the country by lorry.
10. Papers are purchased and read.
11. Papers are put in the recycling bin.

From an ecological point of view the growing and chopping down of timber is going to have the greatest negative impact on wildlife and ecosystems, even in a managed forest like Kielder, during the ten plus years those trees were growing they were sequestering (sucking out) some carbon dioxide from the atmosphere, once felled they do the opposite, they put carbon dioxide into the atmosphere. And while they grew a variety of sub habitats moved in and grew alongside the trees. Smaller shrubby trees, other wildflowers and plants, small mammals, reptiles, birds and insects would all have called the area home.

In deconstructing a tree to turn it into timber and then into paper we are talking about huge industrial processes and a lot of machinery; machines to chop, pulp etc all will be fired up using electricity which might come from fossil fuels.

Other factors to consider are, what chemicals are used in the pulping process and how are they disposed of? Do they for instance end up discharging into a local water course? If so, this could impact a local river ecosystem.

Further factors would concern the inks that are used, many have historically used petrochemicals which have been extracted from somewhere else, the Middle East? Africa? South America? And wherever the oil has been extracted from it will have had an impact on the local environment, often oil deposits sit under forested areas so it will have damaged a local ecosystem.

Once we have the oil derived inks in the factory, they might contain high levels of volatile organic compounds or VOC's, these have been attributed to cancers in workers in the print industry. And again, comes the issue of where wastes from the process are discharged to.

Our love affair with paper as we can see has serious implications for the planet, and it's the scale of the issue that's the problem. This might seem a little crazy, as you read this you are probably thinking, well, just stop. Its really complicated to do that, people are employed right along this chain of production, and to lose the insanity is also to lose people their careers. And that is the thorny place that we reach when looking at the needs of the planet and future people and the needs of people alive today involved in extractive industries who have rent to pay and food to buy. What can we do about this situation? How do we balance the very real needs of people to earn a living, with the needs of the earth to regenerate itself and be viable for future humans to live here?

These are the endless questions that we need to keep asking ourselves until we come up with a liveable solution, we cannot say, 'it won't work, we must just press on regardless of the destruction we create'. But neither can we create millions of unemployed people who used to work in industries that take stuff from the ground or chop stuff down unless we are providing them with other employment opportunities.

Other uses of forest for products that we use, and often take for granted are...

Soya. When grown in the tropics it is often land that was once rainforest. Canada, China and Germany also produce soya and whilst better than tropical rainforest it still suffers the same issues of land use for intensive monoculture farming. South American soy is often used for cattle feed (probably cheaper to buy than soya from the other listed countries), so if you eat meat then it is possible that the soya fed to that cow, or those chickens was grown in an area of what was the Amazon rainforest. It's a horrible thought but a Guardian newspaper investigation discovered that the UK imported a million tons of South American soya in the year 2019, not all of this was checked for origin, so quite literally you could be eating a bit of the Amazon rainforest in your burger bun.

Palm oil. Used in so many products from biscuits to lipsticks, often grown in South East Asian countries like Indonesia and Borneo. Palm oil plantations have a lot of profile in the UK news for the impact on the rainforest and in particular the loss of orangutan habitat which has led to thousands of orangutan deaths every year. Of course, along with the charismatic orangutan are a host of less appealing but nonetheless vital species insects, birds, amphibians, reptiles that also now face extinction. Palm oil production has been linked to the critically endangered or endangered classification of 193 different species worldwide. It is without doubt incredibly harmful to the ecology of the areas in which it grows. Palm oil has become such a significant factor in the human food supply that the loss of habitat to palm oil plantations is about 300 football pitch sized areas of forest AN HOUR. The scale of destruction is staggering. What most people don't realise when they see these vast plantations is that they are monoculture plantations of a plant species that is not native to the area in which it grows. The native plants and their symbiotic creatures; insects, birds, small mammals, reptiles, and large mammals cannot live in a palm oil plantation. Orangutans cannot live in a palm oil plantation. Even if they could, the palm oil will be endlessly sprayed with toxic insecticides and other chemicals which would be harmful. By creating the palm oil plantation for human use you are creating an ecological dead zone.

Most of us have almost certainly eaten a biscuit or a burger in our time. I go through periods of being very switched on, and then some event or crisis occurs, and I fall back on lax ways. It's time consuming to have to check the ingredient list of everything, much better to use Chester Zoo's Palm Oil Scan App which will tell you which brands have turned their back on unsustainable palm oil for good.

CHAPTER FOUR

The ocean ecosystem

We know that about half of the oxygen we breathe comes via photosynthesis that happens in the ocean. It's the same process as in trees, sunlight falls on marine plants like phytoplankton (of which there are trillions, everywhere, in the ocean), on bacteria, on seaweed and other marine plants, which then absorb carbon dioxide from the atmosphere to produce glucose, and just as a tree gives out oxygen as a by- product of this plant nutrient chemical exchange, the oceans release oxygen into the atmosphere.

For us humans to continue to exist on the planet we need oxygen to breathe and food to eat. The ocean provides both. And therefore, if for no other reason whatsoever than its value to humans, we really do need to look after our oceans. They are vital to our continued existence.

When we view the Earth's systems as something that gives value to human life (oxygen to breathe, food to eat) but we don't recognise it as a valuable entity in and of itself ecologists call this Anthropomorphism.

To recognise that the ocean has value as an entity without human interference and that it should be respected as something that has existed for millennia and that all of its systems processes and creatures deserve to exist as they are and not through any relationship to human beings at all, is called Ecocentrism.

The world needs more Ecocentrics.

Oceans are vast and it is impossible to look at them as one single entity, even though each one will flow into the next. As far as ocean ecology is concerned they are separated in a variety of different ecosystems;

1. The inter tidal zone – this is the area around the shoreline, it is where most sea birds will find food. It is subject to tides and so will be terrestrial for parts of the day (when you lie upon its seabed known at low tide as the beach!) At high tide the sea reclaims the area and the waters flood in deluging everything with sea water, this is the time that creatures temporarily marooned in rock pools will take to the seas once more.
2. The estuarine ecosystem, this is where rivers meet the sea, it is usual for there to be a channel in which fresh water will mingle with sea water (ecologists call this brackish water) and this gives way to a unique environment.
3. Salt Marsh, this is an area of land that regularly gets flooded by sea water. Most soils and plants do not do well with salty soil, but some plants and creatures are specifically adapted to thrive in this kind of environment.

4. Mangrove swamps, this is a forest ecosystem that lies in a coastal area and thus the trees are adapted to deal with salt water. Generally found in tropical regions of the world.
5. Coral reef, this is a living exoskeleton, even though it looks like a plant it is actually a living creature. Because coral is vast and remains its whole lifetime in the same place it supports a variety of sea organisms who can live in and around it, small fish, sponges, crustaceans, molluscs and then all of the creatures that feed on them; turtles, sharks and dolphins etc
6. Open ocean, this is the main body of an ocean, it tends to support larger mammals such as whales and dolphins and octopi and sharks who can cover huge distances between feeding zones, are greatly more muscular, and can tolerate the density of this more challenging marine environment.
7. Deep ocean or benthic community, these are the true mysterious creatures of the deep, often extremely weird in appearance (look up Goblin Shark) and adapted to living in a high-pressure environment that is also the darkest and coldest part of the ocean or sea.

Here is a basic ocean food web. We can separate it into trophic levels, trophic is an ecologist term for who's eating whom. And just like forest ecology each organism within an ecosystem is dependent on the other organisms for food, and so when one species dies out or is threatened it will impact the entire ecosystem, and as we know, coral reef ecosystems have been in trouble for many years, and warming in the ocean is the main culprit. When coral dies the hundreds or thousands of creatures that it supported are left at much greater risk of predation because the creatures they ate, lived off the coral, and they now have to leave the safety of the coral to look for food, making them targets for passing dolphins, sharks etc. The coral reef ecosystem is a great example of symbiotic relationships in nature, each creature benefitting from and giving benefit to another creature within that ecosystem. It is also a devastating prospect that these fragile ecosystems are in decline because they are the home of all the small, bright, colourful fish species. To lose the coral is to lose that fantastic wealth of marine gorgeousness.

The Trophic levels of the ocean ecosystems

Trophic level 1 – phytoplankton, the smallest microscopic plants/bacteria that are quite possibly the most important on earth just for their sheer numbers and the fact that most ocean photosynthesis (and our oxygen supply) happens due to their existence.

Trophic level 2 – Zooplankton, tiny little creatures, amoeba, krill, the larvae of fish, squid and lobsters, all feed on phytoplankton.

Trophic level 3 – small forage fish and filter feeders who feed on zooplankton, basking sharks are particularly famous for cruising the ocean with their enormous mouths open filtering up tons of zooplankton as they go.

Trophic level 4 – predatory fish like cod that feed on smaller fish. Marine birds like gulls, gannets, boobies and cormorants. Other marine mammals like seals.

Trophic level 5 – Apex predators, great white shark, orca, whale shark, basically the sea creatures that can take on trophic level 4 for breakfast.

Beyond the animals that we might eat or enjoy watching or photographing and away from the tiny plants that provide 50% of the life-giving oxygen the oceans have several further essential functions that interact with our human needs.

Tides

Tides are driven by gravitational attraction between the earth, the sun and the moon. Every day there are two low tides and two high tides in which nutrients and creatures are re-circulated in the inter-tidal zone. Without these tides the beach area with its rock pools and that hugely diverse area of shallower water would become less diverse and less healthy.

Surface currents

This is where the wind interacts with the surface of the ocean creating friction that in turn creates huge amounts of energy which move the water in what we see as the huge swells that give rise to crashing waves. It is perhaps a surprising fact to know that we have mapped the majority of the earth's surface currents and this is obviously useful for moving cargo ships around the globe, for sailing, even for fishing purposes. By understanding surface currents oceanographers are also able to follow the passage of less desirable elements that end up in our oceans, the waste, pollution and plastic.

If for instance there is a huge oil spill from a tanker in the Atlantic Ocean, scientists will be able to chart the journey of the oil slick and know pretty precisely where it will hit land, and who's ecosystems are locally going to be in for a bashing. This enables preparations to be made to protect what can be protected and to prepare for the many seabirds that will need to be cleaned up after an oil spill.

Chapter Five

The UK loves a fish finger sarnie

I am guilty. I love a fish finger sarnie, I try my hardest not to eat them though, and I honestly cannot remember the last time I did.

So, here's a simple look at the ecological impact of a fish finger, I apologise for the fact that freezing and refrigeration and these energy uses come up at several stages, repetitious but realistic about what brings a finger of fish to your shopping trolley.

1. Fish need to be caught.
2. Fish need to be killed, gutted and chopped into rectangles.
3. Fish need to be refrigerated for transport to a factory.
4. Fish need to be covered in very orange looking breadcrumbs.
5. Fish need to be frozen from this time on.
6. Fish fingers need to be boxed in neat rectangular cardboard.
7. Fish fingers need to be transported frozen to various shops and supermarkets.
8. Fish fingers need to be purchased, cooked, eaten.

As you can see there is nothing really all that simple about a product we feed three year olds, that we consider to be the very baseline of the culinary experience.

1. Fish need to be caught.

I think most of us are aware these days that feeding eight billion people is beyond the capabilities of small-scale fishing vessels. Small scale fishermen are struggling to compete with the new breed of trawler on the block, the supertrawler.

Increasingly the fish we eat might be being caught by supertrawlers, these are enormous boats some nearly 500 ft long, which, to give perspective, is the length of 13 London buses (oceandesk.org) – that dredge the bottom of the ocean. These enormous machines pull nets that are 1km in length and lift everything in their path, resulting in many creatures that are not even suitable as human food to be caught in the nets.

In European waters around the UK a quota system is in force over certain fish species, to prevent overfishing and allow that species time to recover, this happened because of Cod populations becoming dangerously low. When a small fishing boat catches too much Cod it throws the animal back alive.

When a supertrawler catches too many of a certain species or a species that is not human food, it won't be alive, it is thrown back dead. This is the notorious 'by catch'.

Marine animals caught as 'by catch' in nets will end their days being dragged along the seabed, the sheer weight of the nets and animals caught will almost certainly ensure that everything is crushed to death, dolphins, small sharks, even sea birds can end up as by-catch. By the time the fishermen on a supertrawler have sorted the viable product from the excess they are dealing with a lot of death. And these animals will have no further value to the ecosystem or the ecology of the ocean except perhaps as food for some larger fish or seabirds that eat dead prey. If it was a pregnant female, it will never lay its eggs. If it was from a fish species considered endangered like the cod, then it just becomes another tragic loss that drives its species that bit closer to extinction.

By-catch do not be fooled is never just a few unfortunate fish, the WWF (worldwide fund for nature) estimate that 38 million tonnes of by-catch is thrown back dead a year globally. How can that not year on year impact the ecology of the ocean system?

Often fishing lines are left in the ocean that end up snaring marine mammals such as seals, dolphins and turtles, it is estimated that 650.000 marine mammals are ensnared by fishing net every year. Many will end up suffocating because they cannot hunt or swallow food, they die a slow horrible death from malnutrition.

2. Fish need to be killed, gutted and cut into rectangles.

If you are to think like an ecologist, then you have to be aware that nothing in the human food system is without consequence. A huge fishing vessel in which a gutting and filleting and freezing factory is present is millions of pounds of ship. It was built using metals and ores mined from all over the world, and they will have come from areas in which animals have been displaced by that mining industry. It is heated and provides electricity for lighting and refrigerators and freezers on an enormous scale. The staff on board will need to be fed and have a shower and a warm bed after a day of filleting fish in a cold factory. While this part of the fish's journey to becoming a finger sounds relatively low impact, this is industrial food and so it is anything but low impact, huge amounts of oil and gas will be used by the boat to maintain its engines and to keep its function as a fishery, a factory, and a living accommodation.

These boats might stay out at sea for several weeks, all the time consuming more fish and dredging more ocean floor, destroying habitat, lifting ocean plants and leaving nets and other debris that sometimes get tangled around animals.

3. The fish needed to be refrigerated or frozen on their transport from ship to factory for further processing.

Again, fridges, freezers especially industrial scale require huge inputs of energy. Some factories and some transport is going to be carbon neutral, using wind and solar. But sneaking in there somewhere is likely the diesel fuel for the engines which will also power up the refrigeration of that transport. All these things even wind turbines use resources that were mined from some part of the world, and that mining will have had an impact on the ecology of that region.

By now you are probably exhausted by the impact of a small fish finger, like everything else, one little fish finger is a relatively harmless thing, but the scale at which we live means that one little fish finger is part of an industrial system containing millions. And it is scale that will always end up damaging the ecology of the planet. It's nothing we do as individuals, but in buying fish fingers we are not behaving as individuals, we are purchasing a manufactured processed product from a huge food conglomerate.

Of course, you can go fishing by yourself, catch a fish, bring it home, gut it, slice it and cover it in breadcrumbs and fry it, and you will feel very satisfied that you've stuck a big middle finger up to the corporates, and so you should, but before you get too pleased with yourself…

4. Fish needs to be covered in orange looking breadcrumbs.

Wheat. That old chestnut. Let's be honest, we have a long and enduring love affair with wheat, it began with the first recorded farmers in the fertile crescent (modern day Iraq) ten thousand years ago and we've been pretty keen on our daily bread ever since.

There are a lot of political issues around grains; subsidies, scale, equity etc but what you need to know as an ecologist is that wheat is monoculture farming and impactful to the ecology of the region in which it is taking place. Large scale wheat farms need huge inputs of chemicals like fertilizer, pesticides, and herbicides. All these chemicals are damaging to the ecology of the area. All can leach into rivers as run off when it rains. Wheat is the most widely planted monoculture on earth 536 million acres are used each year, an area the size of Greenland, just for wheat.

Pesticides stand accused as evil villain numero uno, because they rarely distinguish between the delightful stripey little pollinator that we all love, the bee, and the irritating and ugly aphid that might suffocate the crop with its very presence. Globally our insects are massively in decline, how often are you truly bothered by flies indoors in the summer these days? I get the odd one. Many of us find insects so disgusting that we think that the loss is not such a bad thing. Ecology however understands their value to the functioning of the environment, and on a farm, soils are made healthy by the contribution of millions of insects and trillions of micro-organisms.

Water another critical indicator for healthy ecology is generally artificially applied to crops by irrigation. The scale of water use is staggering, something like 70% of water humans use in a year is used by the agricultural practices that bring us our food, and wheat is especially thirsty, it takes about 1200 litres of water irrigation per kilo of wheat. Potatoes might only require about 150 litres to give a comparison.

If we were all forced to account for the embedded cost of our lives, that is, the costs of the pollination of plants we eat, the soil health, the cost to the ecology of the planet of extracting minerals and ores, the water inputs etc of every product we buy and use, we would probably all be bankrupt within the hour. As it is, it's the earth herself that accounts for all these costs, we call them 'ecosystem services' it is estimated that the global bill for ecosystem services annually is about $33 trillion. That is, eye watering contributions by bees, other insects, water, sunlight, soil, wind, pollination, ocean currents, and everything else that adds value to human endeavours. In the UK it was estimated that in 2021 our bill for ecosystem services was £47 billion. I wonder how much income tax we'd have to pay were Mother Earth to come asking for her wage cheque.

She is really like the burdened mother who cannot say no to her children, we all know one of those, and we all say, you should just bloody well say no, and she always says, but I love my kids, they are everything to me.

I hope I'm not putting you off your fish fingers but isn't it extraordinary how impactful such an innocent source of protein is?

5. Fish fingers need to be kept frozen from this point on.

Again, freezing is energy intensive, and unless this company is truly avoiding any use of fossil fuels then the energy they will use will come from oil or gas or coal fired power stations, maybe nuclear.

6. The frozen fish fingers need to be boxed in cardboard.

Because we've already looked at the paper industry here are just a few stats that I found by googling cardboard. The UK uses over five million tonnes of cardboard a year. It takes 24 trees to create one tonne. And I don't want to depress you further but once a tree is felled it is no longer a carbon sink but a carbon emitter, so even though in the UK 70% of cardboard is recycled it still isn't absorbing carbon, but yes, it's good to recycle because at least if it can be used more than once fewer trees in total need to be chopped down.

Cardboard can be recycled between 5 and 7 times before the fibres become too weak, so that is some good news. Every time you buy recycled paper or cardboard products you know that no new trees have been chopped down for packaging.

7. Fish fingers need to be taken from warehouse distribution centres to various shops.

I couldn't get an actual figure for how many fish fingers are consumed by the UK population in a single year, but in 2015 according to Birds Eye we ate 1.5 million a day. I wonder if our love of the finger has grown or abated since then? I think that historical figure is still recent enough to be useful for us to think about the scale of consumption. This is only fish fingers, this isn't fish and chips from the chippy down the road, or a posh fish dinner at a swanky restaurant or just fish you purchased from the fish monger that you will steam with lemon, olive oil and a bundle of herbs, no, no, no, this really is just fish fingers, 1.5 million a day. Please never consider a fish finger anything other than a processed industrial mega-scale food from this moment on.

And just imagine that thousands of haulage vehicles every year get filled with fish fingers amongst other frozen products and are driven all over the country stocking the freezers of small shops to huge supermarkets. We could just look at the extraction of petroleum, even though we know that so many other component parts are involved in this scale of transportation.

8. Fish fingers need to be cooked.

Does cooking at home have any direct impact on any ecosystem? Well yes and no. Yes it does because presumably you used gas or electricity to heat your grill? And fossil fuels are mined at great disturbance to the ecosystem from whence they came. However, in this day and age you might be on a tariff that is using mainly wind or tidal or solar and so well done you, you have knocked some carbon off that fish finger that is otherwise lets be honest dripping in fossil fuels.

I hope what the fish finger demonstrates is that everything human beings do has an impact, and even something that seems so insignificant during the course of an ordinary day has a considerable impact on various environments around the world. And whilst it isn't that single act of eating five fish fingers in a bap with tomato ketchup that has created the problem, it is the scale of human activity and the relentless and systematic continuation of that activity that will over time become a problem for the ecology of those environments from which the fish was caught, the boats were built, the fossil fuels were extracted and so on.

Of course, fish reproduce, otherwise the ocean ecosystem would have collapsed decades ago, but what concerns biologists is that our methods of fishing -the huge trawlers and our practices of fishing – the by catch, and our relentless pursuit of more, more, more will push fish to a point at which what they reproduce does not keep up with what is stripped from the ocean. In the waters around Europe practices have changed, breeding seasons are generally observed. Smaller fishing vessels are to be encouraged where small immature fish will be thrown back alive to swim another day. It is up to us as consumers to think about where our fish comes from, how it was caught, whether it is an endangered species. Maybe it is time to cut down on the fish fingers, or at least write to the manufacturers and ask them what practices they use and whether they can give you any reassurances over by-catch, and fishing line caught on marine animals.

When we think ecologically, we have to think at scale, so it is never just about us and our habits it's the fact that we live on a planet with eight billion other humans and every other species of animal that also needs to eat to survive.

Chapter Six

Why does it matter to us if an ecosystem fails?

Across the world there are millions, probably billions of individual ecosystems, they could be a small pond or a large forest, an ocean, a river or a hedgerow. If they are supporting a range of different species and this is their home then they can be said to be an ecosystem, an ecological system that supports the biodiversity (biological diversity) of the planet. And wherever you live, you will have local ecosystems that are contributing to the planet's overall health. And when you consider the enormity of a planet and the miniscule-ness of a village woodland you might conclude that this space, currently under threat of developers isn't such a big loss, that the planets overall ecology can handle the loss of this small area of natural wealth, that even if all the plants, birds, animals and insects of that ecosystem will die as a result of its loss it won't really matter, because the world is a huge vast thing and there are millions of other ecosystems and so the destruction of this loss can easily be absorbed by the greater whole. Right?

And for many years that was true. But it is becoming less and less true. Because it isn't just your local woodland. Its local woodlands to local people right across the entirety of planet earth. Nothing we do in this interconnected globalized world is done in isolation. Everything therefore has an impact on the whole.

And none of us can say which woodland once chopped down will be the one that causes a great tipping point in the ecology of the planet, a moment in which nature unleashes her devastating worst upon us, we presume it will be the Amazon, or some other enormous forest ecosystem, but who knows, it might just as well be the loss year on year of all these smaller less charismatic wild spaces, a field, a wood, a pond all going under concrete to make another housing estate.

And ecologists suspect that we are pushing this variety of life and ecosystem health into what they term a 'tipping point', this is the point at which the systems that regulate plant growth and thus the food we eat could literally begin to fail.

Chapter Seven

But how did we get to this crazy place?

Good question, we know that ecosystems are under threat, and we know that the goods and services we use every day are created using resources that come from all over the globe and this is impacting the ecology of the planet, but why did we allow things to get this crazy, and what are the origins of the madness?

When we look at history there are three identifiable events that might have played a role in our now crazy world of consumerist madness that threatens the very existence long term of life on this planet.

Event One

In the early twentieth century the developing world was carved up by the industrialised nations. Africa especially entered a period know as 'the scramble for Africa' during which the European nations; France, Germany, Belgium, Portugal, Spain, Italy, and the voracious British Empire uninvitedly colonized the continent. Each European nation grabbing as much resource rich land as they could lay their hands on. During this time the people of Africa were treated as second class citizens, enslaved in their own nations whilst the Europeans did their best to strip Africa of its riches, gold, diamonds, precious metals and other mineral resources (for free and using local labour) to ship them back to Europe where they would become expensive manufactured goods.

The two world wars proved very costly for Europe, and this began the end of the colonial period because Europe could no longer afford to keep up the necessary control over the colonies – and control was needed, Africa did not yield willingly, the Europeans with their guns and very likely, their cruelty, were forced upon her.

This also coincided with the rise of the United States as the dominant economic world superpower (noteworthily winning the war with weaponry and aircraft that relied heavily on minerals from Africa and other developing nations).

After the second world war the leaders and chief economists of various countries held meetings (at Bretton Woods) which would decide an economic plan to rebuild the globe in the aftermath of such terrible and expensive conflict. Forty-four nations from across US, Europe, Asia, South America and only four from Africa battled it out to decide how best this was going to happen although the discourse was dominated by the US and the UK.

At these meetings the four Washington Institutions were brought into being that today we know as The

International Monetary Fund, The World Bank, The United Nations and the World Trade Organization.

Poor nations, especially in Africa who were just gaining independence from their European overlords were in a very vulnerable position economically. They were essentially new countries, created by the Europeans and now deserted to their own destiny, without the infrastructure or cash, with which to build their nations.

Agreed at Bretton Woods was that the world bank and the international monetary fund were two bodies who could deliver loans to countries in economic need.

Many African nations rushed to get loans to enable them to begin to build their newly independent countries. What ambitious leader of a poor nation would turn down such an offer? But these loans came with a sting in their tail, huge interest repayments and restrictions on how the money should be used. And so, these countries with large rural populations of predominantly agricultural people found themselves falling back on their loan repayments and having to take further loans to pay the interest on the previous loans and having less and less cash available to pay for advancements such as infrastructure, hospitals, schools, universities, retail, offices, agricultural machinery. And you could argue that this is why Africa has lagged in terms of how it has developed over the past seventy years. As the industrialised nations have amassed enormous wealth it has been impossible for smaller nations still saddled with debts to the world bank and IMF to keep the pace.

It's easy to see that Africa at this time was still largely agricultural and whilst it had huge mineral reserves, it did not have the means of production to turn those mineral reserves into valuable manufactured commodities. African nations had little choice but to import the manufactured commodities from the rich countries in Europe and US and pay many times more, even though the product was created using raw materials mined in Africa. This is one of the huge ironies of what is called neo-colonialism. The rich western world can say that the African nations are sovereign, they have self-determination, they aren't colonies, we aren't doing anything wrong, but to this day we mine or purchase their resources (primary commodities) without ever paying for their true worth. Yes, the fair-trade movement has improved the livelihoods of some farmers of cocoa and tea and coffee, and perhaps that extends into some other commodities like cotton or wine, but have you ever heard of fair-trade oil, or tungsten, or cobalt or diamonds or coltan? These are all referred to in development circles as 'conflict minerals' meaning that their extraction is often in nations with civil strife, the extraction is often corrupt, illegal, unfair, unsafe, and not respecting the ecology of the region it's mined in. Or the lives of the workers in the mines. Or the local people who have been displaced by the mine. And these are all minerals that us westerners take absolutely for granted.

Event Two

This seems like the right place to discuss how petroleum/oil/fossil fuels also played a hand in creating our current lifestyle. This era really was the golden age of cheap and abundant oil, and nobody at this time could have foreseen that this miraculous gloopy black stuff was anything other than manna from heaven. It could quite literally be turned into almost anything as well as fuelling the industrialised world by means of transport, electricity via oil fired power stations and almost every industrial process you can imagine. I have a list of products that are made from oil that I have snitched from The Transition Handbook by Rob Hopkins. They are aspirin, sticky tape, trainer shoes, lycra socks, glue, paints. Varnish, foam mattresses, carpets, nylon or polyester clothing, CD's, DVD's, plastic bottles, contact lenses, hair gel, brushes, toothbrushes, rubber gloves, washing up bowls, electric sockets, electric plugs, shoe polish, furniture wax, computers, printers, candles, bags, coats, bubble wrap, bicycle pumps, fruit juice containers, credit cards, loft insulation, PVC windows, and lipstick. To this list I can add mobile phones, lunchboxes, fleece jumpers and blankets, foundation, moisturizer, mascara, glasses frames, plant milk containers, packaging on food. This list is not exhaustive, because these are just the things that are predominantly made of petrochemicals but what about dashboards and car seats and baby seats, and modem casing and remote controls and, well, the list just goes on. What is important to gain from this is how oil has literally seeped into all our lives even if we don't drive a car or fly or do anything that we recognise to be fossil fuel hungry. And about that label, fossil fuels, well, oil was created hundreds of millions of years ago during what I might imagine (geologists might disagree!) to have been a heating episode on the planet, this turned all those tiny marine creatures and vegetation into this viscous liquid that we recognise today. Million-year events are not available in human lifetimes, so oil is a finite energy source, once it's gone it cannot be replaced.

I think you can see how the extraction of resources, many of which come from poorer nations has been happening over a period of a hundred years, and in order to obtain these minerals, be they ores or oil huge areas of land have been mined at the loss of whatever habitat existed there beforehand.

Event Three

Up until the war goods were purchased on a need only basis. People bought a product and expected it to last a long time. Well, most people, no doubt the super-rich were still showing off, but for the average person like me, goods were purchased only when necessary, and quite possibly they might be once in a lifetime purchases. And they were advertised in a similar fashion. Advertising was functional, here is an example of car advert (from Adam Curtis film The Century of the Self), a funny little car would trundle along a country lane until it came to an abrupt halt outside a train station and out of it would emerge a bald man in a trench coat carrying a briefcase, the vehicle got him to the train on time, bingo! Nothing in this image made the viewer think 'God, I have to have that vehicle, because then I will become a portly middle-aged man with very little hair and a shiny forehead'.

The next significant event that happened once the developed nations had established their economic superiority and enslaved the developing nations into becoming a resource base for the extraction of ores and oil to be turned into manufactured goods – was to turn the world into a planet of rabid consumers.

It was at this point that advertising changed, no longer was it functional, now it was sexy. And what it sold was the American Dream, wherever in the developed world you were you were bombarded by adverts with beautiful women draped across cars, no longer could this vehicle simply get you to the station on time, now it might also get you laid, and by a hot chick. And so, it went on... smiley happy slender attractive families eating, playing, driving and telling you that if you didn't look like this, or drive this car, or eat this bread brand, or have this carpet or three-piece suite... then you were a bit of a loser.

Consumerism gave governments Value Added Tax and made industrialists and retailers ludicrously wealthy. It also justified the endless production that was necessary. Having created enormous car, clothing, and household goods factories there was a constant need to sell the stuff that came off the production line. It wasn't going to work for anyone to say to their workers 'have the year off we aren't going to make anything, we don't want to overshoot on planet Earth's finite resources'. As we have already discussed, the only ones capable of changing that paradigm is us.

Besides, it was providing employment, right? And creating wealth, right? So, we can't just stop, right? Because people would lose their jobs, right?

And this is true and concerning.

But what about finding new, creative less damaging ways of doing things? That will also provide employment. Don't let the big guys off the hook that easily, they have the money and resources to think outside of the box.

If governments and big business were really serious about creating a greener and more sustainable world then they would need to employ more people, to do more things, that's a simple fact. If you were determined to reduce the energy demands of your industry then you would find opportunities along the chain of production into which humans could be slotted to do the work that is currently and increasingly done by machinery, artificial intelligence, and technology. I am not advocating for hard labour, definitely not suggesting any return to that, I am suggesting that we pay creative minds to create new lines of employment through regenerative industries, we need to push for this, every time you choose not to buy a product because of ecological concerns you should email the company and tell them. They need to know we care, and that we will support their products only when we see them making changes in how they do business.

A fairer world is possible. For all of us.

Beyond the hour

Here are some films/podcasts/books that I absolutely recommend you watch, listen to or read bits of. I have looked-for easy-going sources, because I know you're busy and time is short.

Chapter 1 What is ecology?

For a deeper dive that isn't too heavy I cannot recommend this book enough,

Impacts of Human Population on Wildlife – a British perspective by Trevor J.C.Beebee. Read even if you aren't a UK based ecologist. I read books about US ecology and I'm not planning to go to the US, but I enjoy them and learn from them, ecology is ecology, and whilst this is an academic book, its actually an easy read on a difficult subject.

The Blindboy podcast (available from your favourite podcast site) has many pods on biodiversity and conservation. If you don't like your ecology littered with expletives, maybe it's not for you but I suspect that if you're a bit of a lad, or a ladette, this lads style might be just right for you.

- Mental health, biodiversity and mythology Mar 29th 2023
- Speaking to an expert about climate change and biodiversity Nov 15th 2023 (says brilliantly everything I might have failed to say brilliantly)
- Venom, bites, snakes, scorpions, pain Mar 13th 2024 (for introduced exotics this a must hear)

How wolves change rivers, it's a beautiful film, explains exquisitely how everything in an ecosystem has its place, and how the ecology is affected by the loss of a keystone species which leads to the over-population of another species. Its shot in Yellowstone national park so the nature and sceneries are beautiful, I have watched this short many times and it always brings a tear to my eye, and just a feeling that our world is beautiful, we, the little people must save it. (4.5 mins)

Chapter 2 Ecology of the forest/woodland

Some short films from YouTube that might inspire:

Scotlands Native Woodland from Forestry Commission Scotland and starring Nick Baker. (16.50 mins) Just a lovely and informative introduction to woodland ecology. I don't live in Scotland but just love a wander in the woods and hearing about how many different forests and woodland types there are, and that I most likely walk my dog in a lowland mixed deciduous woodland and that this type of woodland is really old makes it all the more enjoyable.

The secret history of dirt by Vox. A really informative short film on everything you need to know to grow your love and appreciation of the soil beneath your feet. (at 10 minutes 38 seconds)

Living Soil is an hour-long film from the Soil Health Institute of America. An hour is a big ask, but if you feel like it you will learn how the US destroyed its soil in the 1930's from monoculture overcropping and endless tilling, this was known as the dustbowl and caused food shortages and famine (if you've ever read The Grapes of Wrath by John Steinbeck you will know it was a really miserable moment in US history). Its really worth the watch, but if you're short on time here is a section about how modern farmers are adopting no tilling (ploughing over the topsoil) methods to make their soils healthy and rich and full of nutrients. (minutes 27.39 – 33.18)

Fantastic Fungi is THE film that will make you fall in love with mycelium. It was on Netflix the last time I had a Netflix account.

Chapter 3 To trade of not to trade

The story of stuff is a must-see film from 2007 from the wonderful Annie Leonard about how the extraction/manufacture/trade in consumables has impacted the ecology of the world. Its 21 minutes long and worth every minute.

And if you like it, she has made a load more videos available on YouTube and they are all fantastic.

Chapter 4 How we have impacted a woodland/forest

The problem with palm oil from Fight for the Forest. This short film highlights the year-on-year impacts of palm oil production worldwide. It's an upsetting and unsettling 2 minutes and 46 seconds.

Conflict minerals, rebels and child soldiers in Congo from Vice.

Phones and laptops = batteries = coltan -= conflict minerals. Its 38 minutes long and makes me want to live on a remote island and give up communicating…with anyone.

Poison Fire is about oil extraction in Nigeria, another disturbing film about the nature of how the rich world goes into poorer nations, takes the resource, and destroys human and ecosystem health. 90 mins is long but still a short amount of time in which to live and breathe the life of real actual humans living with the effects of the petrol you filled your tank with this morning. Five minutes though is better than nothing.

Poison Fire is 15 years old, so I thought I'd look up the situation in the Niger Delta today (2024) – I was expecting to find it a clean, green beautiful place after all the oil company responsible for destroying it is worth billions. Here was an article I found from the Guardian newspaper from February 2024 Shell must clean up pollution before it leaves Niger delta, report says | Pollution | The Guardian

The true cost of mining for the $500 billion electric car industry – by Business Insider is a brilliant must watch. I am not having a go at electric cars. I wish I could afford one. As someone who lives in a hilly city, I love it when an EV cruises past me and I don't have to fill my lungs with noxious car fumes. This is an amazing film for its thorough and honest appraisal of the mining industry globally. It's all in there, from the highest tech plants in US to child labour artisanal mines in Democratic Republic of Congo. I think as rabid consumers we all need to watch more films like this.

Chapter 5 The ocean ecosystem

My Octopus Teacher is my favourite film about something that lives in an ocean. That little octopus was enough to convince me that Octopi are intelligent sentient beautiful creatures, I just adored her. The trailer is 2.5 minutes, but I recommend the full film on Netflix.

Chapter 6 But does it matter if we lose an ecosystem?

Saving ancient woodland from HS2 destruction from Stop HS2 (4 mins). I'm not sure how much of this ancient woodland still stands? And I wonder how many creatures died in the felling. What I love in this short film is that local people got involved and set up camp in the woods to oppose the destruction. Direct democracy, perhaps it didn't work but at least they tried.

Chapter 7 How did we get to this crazy place?

<u>Africa states of independence</u> – the scramble for Africa by Al Jazeera such an accessible and necessary film. (45 mins)

<u>The Impact of Colonialism a century later – by BBC Select (trailer 2 mins) I haven't been able to find the whole film, come on BBC! The trailer is powerful indeed.</u>

<u>Petroleum, modern history of oil on a map</u> by GeoHistory is a classroom style history of the world of crude. (14 mins)

<u>The Century of Self</u> by Adam Curtis is just mind blowing, watch it and be appalled at how we have all been duped to line the pockets of the one per cent. Please watch. I find it incredibly frustrating that films like this aren't on the national curriculum just so kids emerge from school knowing that we became part of an economic plan that said, to have a sense of value and self-respect in this society you must be a rabid consumer. It doesn't have to be this way. Part One is approx. an hour.

What next...

Reading this kind of material can feel disempowering, all is not well in the world, and you aren't quite sure how you can respond to the gravity of the situation. Sometimes you can just feel guilty for being a western human, living a western human existence. That's why so many people don't engage at all. The Guardian newspaper reported on a survey that found that most people in the UK don't even understand the meaning of sustainable or green when applied to products that they might buy.

https://www.theguardian.com/environment/2024/jan/24/key-climate-language-poorly-understood-by-majority-in-uk-poll-finds

This is serious disengagement, and I suspect it comes from people feeling that it will be expensive to go green, or that they already have a life that is stressful enough, this is just another layer of difficulty for them to navigate.

Here are some quick definitions to make you feel less stressed out.

The term sustainable was first coined by Gro Harlem Brundtland the then Norwegian Prime Minister at the World Commission on Environment and Development in 1987, and it set out these aims, <u>sustainable means satisfying present needs without compromising future generations to meet their needs</u>.

Put more simply it means that what we have today shouldn't come at the expense of future humans being able to expect to have the same.

The problem is that the western lifestyle we have today is generated using vast quantities of resources like oil, coal and gas and minerals like coltan, tungsten, cobalt, iron, copper, and radionuclides like uranium.

All of these are non-renewable or finite resources. This means that the earth cannot re-create these resources, once they're gone, they're gone.

When a product is labelled sustainable it means that the processes that were used in its production are usually using renewable resources. Take palm oil, sustainable palm oil will be grown in an area where it did not cut down an ancient rainforest and threaten with extinction a species like orangutan. So, it is worthwhile to look for a sustainable option if its available.

If a product is labelled green it means that it does not contain chemicals (or has minimal chemicals) harmful to human or ecosystem health. Its ingredients have been sourced in as sustainable a way as is possible, so no soya from the Amazon rainforest. Where possible in the chain of its production, recycling of materials, say the bottle, will be in place.

What next...

1. Love nature. The park will do. It's worth asking your local council what they're doing to encourage wildlife? And how you can get involved to make the place you live more natural. Are they planting native flowers and trees? Do they have to mow all the green areas? Could they plant some edible trees, nuts, fruit etc? Identify the birds or butterflies or beetles, join a group (or create a group), start making lists of what you see, learn to identify species of plants. It's a free and easy way to enjoy the natural world.
2. Buy some native wildflower seeds and sprinkle them as you go about your day.
3. Join a local wildlife trust, or the RSPB, or Friends of the Earth all of these organizations will release to the press sudden increases in membership, this sends messages to the government that 'the people' are concerned about the state of the planet and it informs decisions, especially if they have some destructive road building project pending that is going to carve its way through important wildlife areas, its really up to us to let them know we care about the ecology of this planet.
4. Invite friends over to watch films about the environment, Top Documentary Films has a good

environment section, work through it. I used to invite mum's round when the kids were at school, inevitably we'd all end up gossiping about something else, but we watched the film first and tried to set aside twenty minutes for discussion, but we always ended up on the usual school mum scandals – oh we were such average eco warriors!
5. Start an ecology book club, the more clued up you are, the less you feel good about doing anything that harms the environment, the more you feel better about being skint. I live by these principles; I have realised that being a generally skint person keeps me from being a planet destroying person. I like shiny things and I'm about as greedy as the next person, so I'm quite glad that I can't afford to be more of a rabid consumer.
6. Try and eat as much local food as possible.
7. Try to cut down on plastic packaging, start with one item a week. Bring your excess packaging back to the shop and ask them to recycle it.
8. Get on an allotment waiting list and set about growing some of your own food.
9. If growing stuff appeals, then learn about Permaculture a method of growing food that is very ecologically minded.
10. Buy as much of your stuff second hand as possible.
11. Ask manufacturers if they use recycled packaging, sustainable palm oil, non-Amazon soya, fewer harmful chemicals, every email you send to a corporation matters, they audit everything, so if enough people are asking these questions, they know their customers are concerned, they will respond, because they want your cash.

12. Try to cut down or give up meat, and dairy, it takes so much land and water to grow food for animals – approximately 40% of land used in agriculture is growing food for cows/chicken/pigs/sheep, and so the fewer animals the more land could be used to grow food for humans and to just exist for nature.
13. Try to fly less.
14. Try to drive less.
15. Try to use public transport. Write to rail companies and ask them why in an ecological crisis they aren't offering more cheap fares?

References

Hewitt. T (2003) *Half a Century of Development* in Allen.T, Thomas. A. Poverty and Development into the 21st Century, pp. 289–294, Oxford University Press

Morris, D, Freeland, J, Hinchcliffe S, Smith S (2003) Changing Environments, pp. 151–158 *Below the surface: land and soil.* Wiley Press in association with the Open University

Morris, D, Freeland, J, Hinchcliffe S, Smith S (2003), Changing Environments, pp. 207–215 *Oceans,* Wiley Press in association with the Open University

Hopkins, R (2008) The Transition Handbook – from oil dependency to local resilience, p. 18 *A few of the things in our house made from oil.* Green Books

Devall. B, Sessions. G. *Deep Ecology* in Thinking through the Environment, 1999, Routledge

Callahan. Daniel. *What obligations do we have to future generations* in Thinking through the Environment, 1999, Routledge

Milman, Oliver. The Insect Crisis, 2023, Atlantic

UK: packaging waste generated & recycled by material | Statista

www.ingramcontent.com/pod-product-compliance
Ingram Content Group UK Ltd.
Pitfield, Milton Keynes, MK11 3LW, UK
UKHW040703280225
455691UK00001B/8